JOHN DEERE
TRACTORS
A PHOTOGRAPHIC HISTORY

Andrew Morland

Voyageur Press

First published in 2008 Voyageur Press, an imprint of MBI Publishing Company LLC, 400 First Avenue North, Suite 300, Minneapolis, MN 55401 USA

Voyageur Press titles are also available at discounts in bulk quantity for industrial or sales-promotional use. For details write to Special Sales Manager at MBI Publishing Company, First Avenue North, Suite 300, Minneapolis, MN 55401, USA.

To find out more about our books, join us online at www.voyageurpress.com.

Library of Congress Cataloging-in-Publication Data

Morland, Andrew, 1947-
    Legendary John Deere Tractors/by Andrew Morland.
        p. cm.
        ISBN-13: 978-0-7603-3293-1 (hardbound w/ jacket)
    ISBN-10: 0-7603-3293-2 (hardbound w/ jacket) 1. John Deere tractors—History.   2. John Deere tractors—Pictorial works. I. Title.
TL233.6.J64M675 2008
629.225'2—dc22

Editor: Leah Noel
Designer: Mandy Iverson

3 1561 00216 4055

Printed in China

On the frontis:
1970s-era John Deere logo
In 1968, the John Deere logo was overhauled, getting a new contemporary, clean-cut look. The design was streamlined to show a deer's straight-side silhouette with just two legs, instead of the four, and one four-point rack of antlers. According to a company memo, the new logo provided "better reproduction and greater readability under a wider range of usage."

On the title pages:
Model 4020 Diesel High-Crop
The 4020 was the most popular tractor of the 1960s, now considered by many to be an all-time classic Deere tractor. Its design, build, and reliability were all solid.

On the contents page:
1935 Model B
The cast-iron John Deere–badged radiator and overhead steering post, for the Model B, the company's best-selling two-cylinder tractor.

On the acknowledgments page:
1967 Model 3020 Row-Crop
Almost immediately after the introduction of the 10 Series, models from the 20 Series began to replace the initial New Generation tractors. The 8020 came out in 1961, and the 3020 (pictured here) rolled off production lines in 1963.

Contents

Acknowledgments

Many thanks to the owners and restorers who gave up their valuable time and allowed me to photograph their historic tractors and, in many cases, helped me find other collectors in their neighborhoods and farther afield. This book would not have been possible without them and all their kind help. They include the following: Tony Adams, Kenneth Anderson, Neil Bailey, Mike Bastin, Kent Bates, Stephen Batten, Daniel Binet, Pierre Bouillé, Glen Braun, Don Bray, Arvin Busenitz, Robert Coles, Bruce Copper, Robert Couch, Chris Dart, John Davis, Rick Van Esser, Malcolm Godard, Mark Farwell, Robert Fry, Ivor Grant, Peter Hawkin, Robert Heckendorf, Denzil Horler, Maurice Horn, Nigel Hutchings, Brent Johnson, Neal Jackson, Kenneth Kass, the Keller family, Kings Farm Machinery, Paul Kleiber, Jaak Langers, the Layher family, Larry and Melanie Maasdam, Don Macmillan, Marv Mathiowetz, Don Miller, Lyle Pals, Kelsi and Duane Ver Ploeg, Bob Pollock, Rich Ramminger, Tony Ridgeway, Aubrey Sanders, Donald and Ruth Schaefer, Harold Schultz, John Smith, and Don Wolf.

Thanks also goes to the farming contractors whose machines I photographed, including Tony Curtis, Malcolm Harding, Dennis Clothier, and Gray and Sons. In addition, Peter Love, formerly of *Tractor & Machinery* magazine, and Rory Day at *Classic Tractor Magazine* were very helpful.

More thanks goes to the John Deere Two-Cylinder Club and the John Deere historic site at Grand Detour, Illinois; the Northern Illinois Steam Power Club of Sycamore, Illinois; the Great Dorset Steam Fair; the Somerset Steam Spectacular and Country Fair in Low Ham, Langport; South Somerset Agricultural Preservation Club's *Yesterday's Farming*; and Mark Farwell's *Dorset Tractor Working Day*.

A special thank you to John Deere and Steve Mitchell of ASM Public Relations Ltd., who is a shining example of how PR should work!

Thanks also to Walter and Bruce Keller for all their help over the last 25 years or so. I have seen their collection grow into one of the best and biggest collections of John Deere tractors in the world.

Finally, eternal gratitude to Bob Pripps—friend, tractor expert, and author—for his excellent company over the last 25 years and for making going on the road around the United States so much fun.

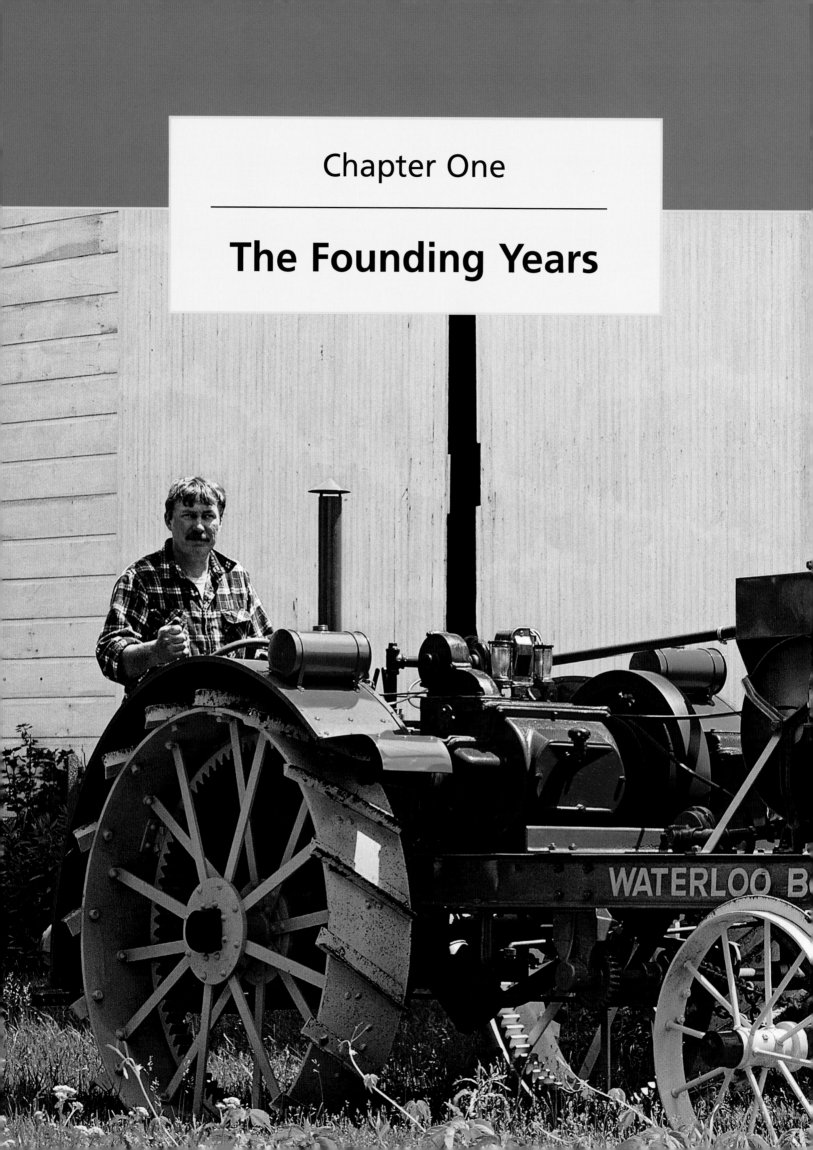

Chapter One

The Founding Years

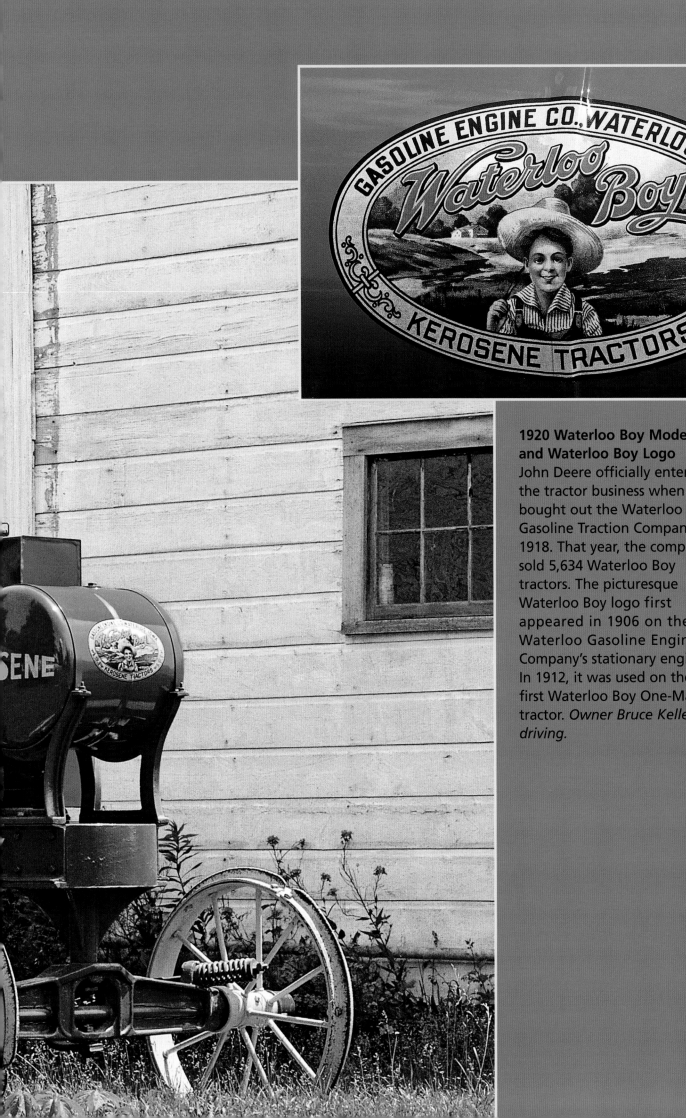

**1920 Waterloo Boy Model N
and Waterloo Boy Logo**
John Deere officially entered
the tractor business when it
bought out the Waterloo
Gasoline Traction Company in
1918. That year, the company
sold 5,634 Waterloo Boy
tractors. The picturesque
Waterloo Boy logo first
appeared in 1906 on the
Waterloo Gasoline Engine
Company's stationary engines.
In 1912, it was used on the
first Waterloo Boy One-Man
tractor. *Owner Bruce Keller
driving.*

9

1892 Froelich Replica
In 1892, Iowa thresherman John Froelich used a single-cylinder 2,155-cubic-inch Van Duzen gasoline engine mounted to a chassis to help harvest fields in South Dakota. The invention is now considered the first successful gas tractor. This replica was built by workers at Deere's Waterloo, Iowa, works for use in a documentary film in 1937. It is on display at Deere & Company's main headquarters in Moline, Illinois.

1917 Dain All-Wheel-Drive

In 1911, Deere & Company purchased Joseph Dain's haying equipment company, then based in Ottumwa, Iowa. Dain stayed on at Deere and was eventually given the job of developing the company's first tractor. The prototype for the Dain all-wheel-drive three-wheeler appeared in 1915. The Deere board wanted the tractor to be priced at $700, but its development costs made that impossible.

1917 Dain All-Wheel-Drive

This John Deere Dain all-wheel-drive tractor is one of only two complete tractors that remain from a production of 100. A third survivor does exist, but most parts are missing. Production of these tractors ceased with the death of Joseph Dain. *Owner: Northern Illinois Steam Power Club.*

1917 Dain All-Wheel-Drive
Walter McVicker designed the Dain all-wheel-drive's four-cylinder engine with a bore and stroke of 4.50x6.00 inches (112.5x150mm). It produced 24 belt horsepower.

1917 Dain All-Wheel-Drive
The gear lever for the two-speed forward and two-speed reverse transmission with friction drive similar to a Model T's.

1917 Dain All-Wheel-Drive
At right, the chain drive to the two front wheels. Above, the chain drive to the single rear wheel. Note adjustments for chain wear, which was considerable in muddy conditions.

Waterloo Boy Model R

After inventing his self-propelled gas engine tractor, Joseph Froelich founded the Waterloo Gasoline Traction Company in late 1892. The company later decided to concentrate on producing stationary gas engines, dropping the word traction from its name. But by 1914, the company was back in the tractor business when it built the Waterloo Boy Model R. *Owner: Tony Ridgeway.*

Waterloo Boy Model R
Between 1917 and 1918, Waterloo Boy Model Rs had a two-cylinder 465-cubic-inch engine that produced 25 belt horsepower and 16 drawbar horsepower. An earlier R engine with 396 cubic inches of displacement produced only 12 horsepower at the drawbar. *Owner: Tony Ridgeway.*

Waterloo Boy Model R
The drive gear on the Waterloo Boy Model R is much smaller than the one on the Model N. The R's has only one forward and reverse gear.

1920 Waterloo Boy Model N
The Waterloo Boy Model N was introduced in late 1916. The Model N's two-cylinder kerosene engine, with 465 cubic inches of displacement, ran at 750 rpm, putting out 25 belt horsepower. This early 1920 version was the last featuring a low fuel tank position and chain wind-less steering. *Owners: The Keller family.*

1920 Waterloo Boy Model N
Right, the spring steering damper on front axle for the chain steering.

1917 Overtime Model R

During World War I, with German submarines threatening the imported British food supply, home food production had to be increased. As part of that effort, about 4,000 unassembled Waterloo Boy Model Rs and Ns were shipped to London. They were put together by L. J. Martin's Overtime Farm Tractor Company and painted in the firm's colors: gray, green, and orange. The power for these machines came from a two-cylinder overhead-valve 465-cubic-inch engine rated at 12 drawbar horsepower and 25 belt horsepower. The Overtime was advertised as the "national food production champion" of England and Wales. *Owner: Mike Bastin.*

Overtime Logo
The Overtime logo featured an image of a clock, with hands pointing to either 11 o'clock or 1 o'clock.

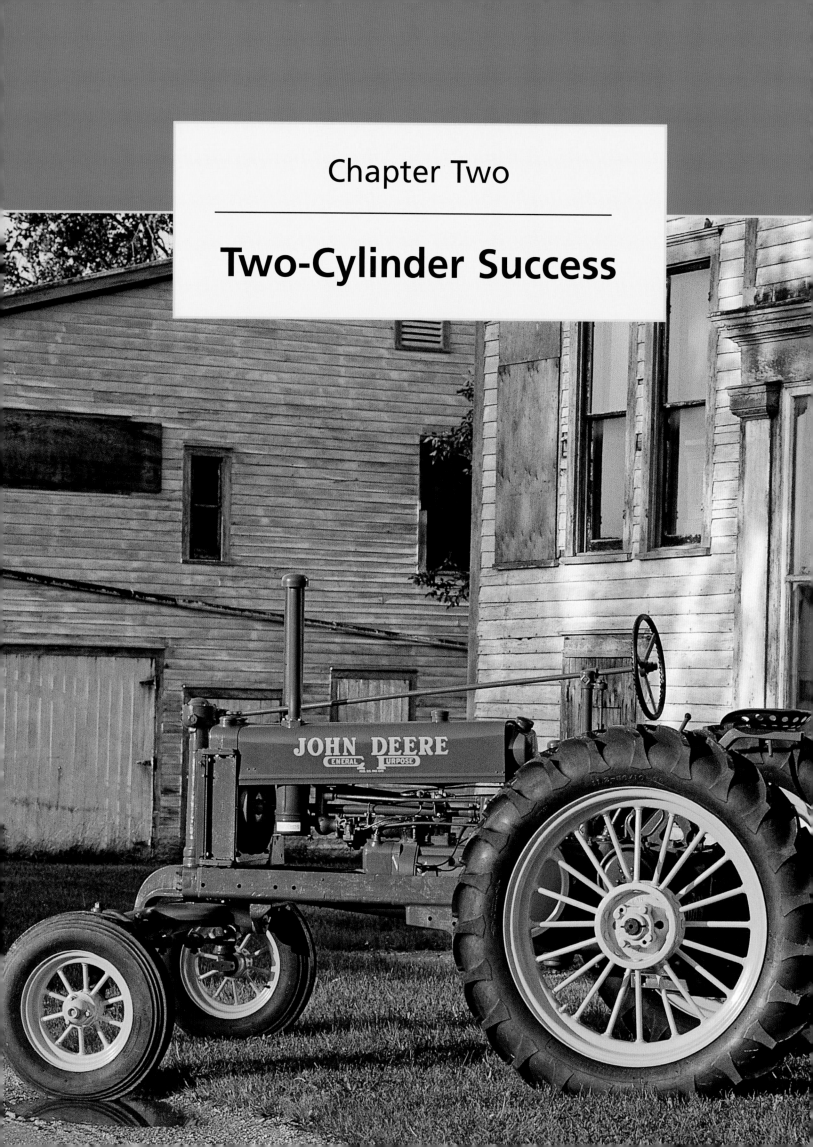

Chapter Two

Two-Cylinder Success

1935 Model BW
This Model BW is one of the many different variations Deere made of its popular two-cylinder Model B tractor. These different versions were designed to easily navigate through many different crop setups. The Model B tractors introduced in 1934 became the most popular of all of Deere's two-cylinder tractors. *Owners: The Keller family.*

Model D

The Model D, originally conceived by the Waterloo Gasoline Engine Company, was the first John Deere–branded tractor. The standard-tread three-plow machine was introduced in 1923 and sold alongside the Waterloo Boy until the Waterloo Boy's production ended in 1924. *Owners: The Layher family.*

1925 Model D
The spade lugs in the tractor's steel 46x12-inch wheels helped the 4,000-pound Model D to keep moving in the field. With a two-speed transmission, the tractor could go up to 3.25 miles per hour.

1925 Model D
The Model D's engine was a two-cylinder overhead-valve 465-cubic-inch powerplant. It produced 30.40 belt horsepower at 800 rpm. The Model D's magneto ignition was mounted high up on the engine. The tractor had a lubrication specification diagram mounted on the end of the fuel tank in front of the driver.

1925 Model D Spoker

These spoked flywheel Model Ds later became known as "the spokers." The 1923 models had 26-inch spoked flywheels, but in October 1924 Deere & Company switched to 24-inch spoked flywheels. The first solid flywheel models were sold in January 1926.

1925 Model D Spoker

Above, Don Macmillan on his Model D Spoker. Don is a well-respected author of more than 10 books and the authority on Deere & Company. He was appointed the first Deere dealer in Great Britain in 1958 and became an enthusiastic collector of Deere tractors.

1935 Model D

By 1935, the Model D weighed a hefty 5,270 pounds and had a three-speed transmission. The tractor was powered by a two-cylinder overhead 501-cubic-inch engine rated at 42 belt horsepower and 30 drawbar horsepower. This Model D is fitted with the 13.5/28 rubber rear tires and 7.50/18 rubber front tires. *Owner: Don Wolf.*

Model C

The Model C was John Deere's answer to the popular Farmall tractor introduced by International Harvester in 1924. Launched in 1928, the Model C was an all-purpose tractor, designed for plowing, pulling, and belt work, as well as cultivating and planting. The Model C had many field problems: 53 out of the 99 tractors that first came off factory lines had to be rebuilt. *Owners: The Layher family.*

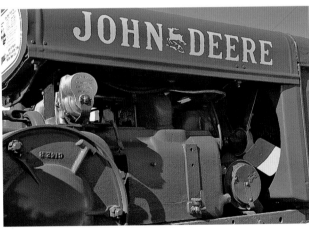

Model C
The Model C was nearly two thirds of the power and weight of the Model D. Its two-cylinder L-head 312-cubic-inch engine produced 10 drawbar horsepower and 20 belt horsepower.

Model C
The Model C, showing the straight-cut round-end exhaust pipe and brass carburetor with cylindrical air filter. The air filter's position between the radiator and cylinder head reduced the tractor's performance and even shortened engine life because of the amount of dirt blocking the air intake.

1929 Model GP

This rare side-steer tricycle, serial number 204213, was one of only 23 built between 1929 and 1930. Note the housing for the chain final drive, which was popular in the early 1930s. *Owners: The Keller family.*

1930 Model GP

This Model GP has a 312-cubic-inch engine that produced 20 horsepower at the belt. Because of its lack of power (and the fact the Model GP was $200 more expensive than the Farmall equivalent), the engine was upgraded in late 1930 by increasing its bore from 5.75 inches to 6 inches. The air filter was also moved to a stack above the hood to get more power. As a result, the GP was rated at 16 drawbar horsepower and 24 belt horsepower. *Owner: Don Wolf.*

1932 Model GP
This 1932 Model GP, serial number 228863, is one of only 385 built in 1932. The attached implement is a Model 301 check-row planter, made between 1928 and 1935. In 1932, rubber tires became an option on GPs. *Owners: The Keller family.*

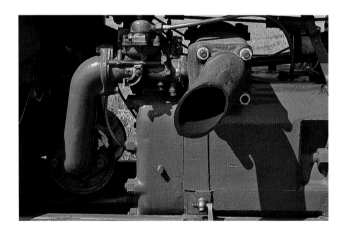

1930 Model GP
The exhaust, carburetor, and air filter on the L-head side-valve two-cylinder engine.

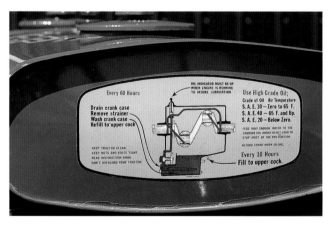

1930 Model GP
Service information on the back of the fuel tank, in full view of the driver.

1933 Model GPWT

The Model GPWT (general purpose wide tread) was built for only one year, with 444 coming off factory lines. This one is serial number 405135. Note the transverse implement mounting holes behind the front wheels. At bottom right is the GPWT's overhead steering mechanism, similar to the one used on the successful Farmall and Oliver tractors. *Owners: The Keller family.*

1938 Model ANH

This tractor had a single front wheel and stood higher than the standard Model A. It is one of only 26 produced. *Owners: The Keller family.*

1934 Model AW

Deere & Company introduced the Model A, intended for crop work, in 1934. Soon after came the Model B. Both were eventually produced in many different variations (the Model A had 14), each designed to meet diverse farming needs based on farm size, geography, and crop. This is the Model AW (wide front). It came with a two-cylinder 309-cubic-inch engine that put out 18.72 drawbar horsepower and 24.71 belt horsepower at 975 rpm. The transmission had four forward gears. *Owner: Don Wolf.*

Model AOS
This Model AOS (orchard streamlined) was built on January 21, 1937. It is one of only two AOS models with an engine oil dipstick. Fewer than 900 Model AOs were built between 1936 and 1940. *Owners: The Keller family.*

1946 Model AO
This is the low-slung orchard version of the Model A. Its engine was rated at 26.30 belt horsepower and 20.35 drawbar horsepower.
Owner: Betty Norton.

1947 Model AR

In 1941, Deere updated the Model AR, which was first introduced in 1935. The tractor now came equipped with a 321-cubic-inch engine that produced 18.7 drawbar horsepower and 24.7 belt horsepower on kerosene. When using gasoline, the engine could put out a hefty 34.9 horsepower at the drawbar and 39.1 horsepower at the belt. The Model AR became the last John Deere tractor to be styled. Before 1948, none of the ARs like this one had model decals. Between 1935 and 1953, around 30,000 ARs were manufactured. *Owner: Neal Jackson.*

1935 Model B

The Model B was two-thirds the size of the Model A and was designed for the small row-crop farmer looking to replace his last team of horses. The two-plow tractor came with a four-speed transmission and PTO. Farmers had the option of equipping the B with either steel wheels or rubber tires. The tractor's two-cylinder 149-cubic-inch engine had a bore and stroke of 4.25x5.25 inches. At the 1935 Nebraska tests, the Model B was rated at 11.84 drawbar horsepower and 16.01 belt horsepower at 1,150 rpm.

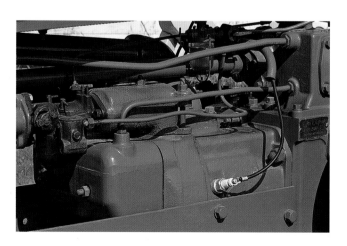

1935 Model B
The brass tag on this Model B sports the serial number B1000, indicating that this tractor is the first of the 300,000 built by John Deere.
Owners: The Keller family.

1936 Model B
These early row-crop tractors had the John Deere name cast into the rear axle.

1935 Model BN

The Model BN was designed especially for market garden and vegetable growers, with a single 22.5x8-inch front wheel, a narrow adjustable rear axle, and rubber tires. *Owners: The Keller family.*

1935 Model BW

This Model BW (wide front) has the F&H spoked wheels. Its 149-cubic-inch (2,441cc) two-cylinder engine, rated at 16.01 belt horsepower, is mated to a four-speed gearbox. *Owners: The Keller family.*

1935 Model BR

This rare tractor is one of the lowboy Model Bs that was never built with Henry Dreyfuss' styling cues. *Owner: John Davis.*

1936 Model B

Owner: Don Wolf.

1938 Model BWH

The high-clearance Model BWH is highly collectible because only 51 came off factory lines. The BWH offered wide wheel spacing (from 56 to 104 inches in the rear) for cultivation of corn, cotton, and sugar cane. *Owners: The Keller family.*

1935 Model B

Below, the cast-iron John Deere–badged radiator and the overhead steering post and gear set.

1938 Model BWH
Note the front and rear axles, extended in their wide position.

General Purpose Logo
The John Deere trademark on Model A and Model B tractors included the words "general purpose" to emphasize that these tractors had similar capabilities to International Harvester's Farmall.

1945 Model BO Lindeman

To make crawler tractors for apple orchards and market gardens, John Deere shipped Model Bs to Jesse Lindeman in Yakima, Washington, who outfitted them with tracks and a running gear. Lindeman and his brothers built 2,000 Model B crawlers before the Model M was announced. In 1946, Deere & Company bought out Lindeman, but kept him on to help convert the M into the MC crawler. In 1954, Deere moved the crawler operations and many of the employees to Dubuque, Iowa. *Owner: Harold Schultz.*

1945 Model BO Lindeman Crawler
Below, special angled levers for controlling steering, clutches, and brakes.

1937 Model 62
John Deere's Model 62—developed from the Model Y at the Moline Wagon Works—was the smallest John Deere tractor made before World War II. Production began in September 1937, with this tractor being the first off factory lines. Only 79 Model 62s were built between January and July of that year. The tractor was powered by a 56.5-cubic-inch two-cylinder gasoline engine built by Hercules. The engine had a bore and stroke of 3x4 inches. *Owners: The Keller family.*

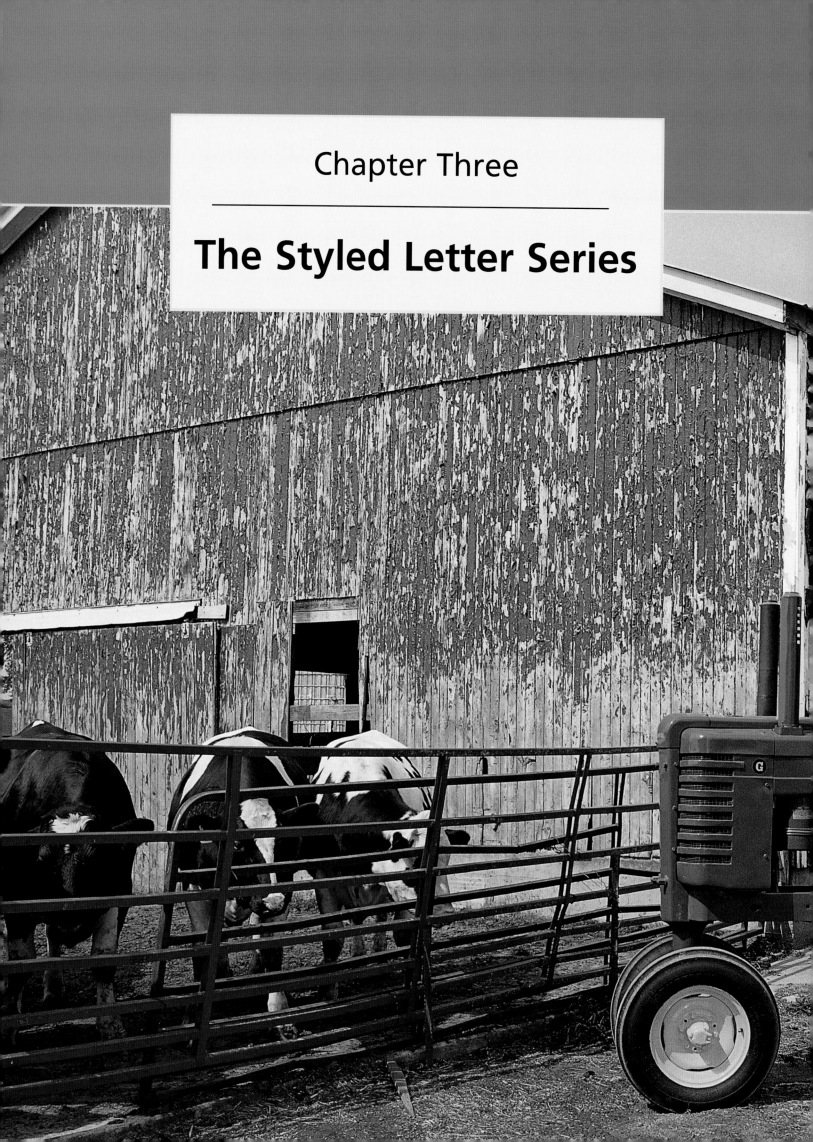

Chapter Three

The Styled Letter Series

**1951 Model G and
1949 Model AWH Styled**
Beginning in 1937, the tractors
coming out of Deere & Company
factories had a new look, thanks
to industrial designer Henry
Dreyfuss. Dreyfuss streamlined
the tractors with the goal of
improving their appearance,
utility and safety, ease of
maintenance, cost of production,
and sales appeal. The Model G
featured bulging chassis frame
side rails to make space for its
massive two-cylinder 412.5-cubic-
inch engine with a bore and
stroke of 6.125x7 inches. *Owner:
Lyle Pals.*

1949 Model AWH Styled
This is a 1949 Model AWH, with a Dreyfuss-designed pressed steel frame. Its 321-cubic-inch engine—producing 34.14 horsepower at the drawbar and 38.02 horsepower at the belt—is mated to a six-speed gearbox. *Owner: Lyle Pals.*

1952 Model AWH
This rare wide-front high-crop tractor has a two-cylinder (5,258cc) engine that puts out 39.1 horsepower on the belt. This model was produced in the last year of the late-styled Model As. The Model 60 was introduced in 1953. *Owner: Larry Maasdam.*

1950 Model AR
The Model AR was the last of the Deere tractors to be given the modern Dreyfuss look. Note that the replacement mufflers on this tractor are longer than the originals. On most ARs, the mufflers are the same length as the air intake. *Owner: John Davis.*

1940 Model BWH-40
This extremely rare version of the Model B has high clearance and a wide front end, but its axles also can narrow down to straddle 40-inch rows due to its specially shortened axle housings. Deere records show only six BW-40s were built. *Owners: The Layher family.*

Model BW
This Model B with a wide front end has unusual British-built wheels. Here, it is cultivating with rear- and mid-mounted implements at Fovant, in Wiltshire, Great Britain.

1952 Late-Styled BWH
This late-styled Model BWH is from the last year Model Bs were produced and has a pressed steel frame and a 190-cubic-inch engine. The Model B was the best-selling two-cylinder Deere tractor, with more than 300,000 built. *Owner: Lyle Pals.*

1948 Model D
The Model D had a more subtle Dreyfuss styling. Its two-cylinder 501-cubic-inch (8.2-liter) engine put out 38 drawbar horsepower and 42.1 belt horsepower at 900 rpm. The powerful tractor had no trouble pulling in top gear (with a top speed of 5.25 miles per hour) while light cultivating. *Owner: Malcolm Godard.*

1947 Model G
The powerful row-crop Model G was marketed as a heavy-duty three-plow tractor equipped with a 412.5-cubic-inch engine. The powerplant produced 34.49 horsepower at the drawbar and 38.10 horsepower at the pto/belt. Rubber tires became standard on the G in 1941.

1953 Model G High-Crop
The Model G was produced from 1938 to 1953, with 64,000 rolling off factory lines. *Owner: Maurice Horn.*

Model H

The styled Model H row-crop tractor was manufactured between 1939 and 1947, and approximately 60,000 were produced. As shown here, many were equipped with mid- and rear-mounted cultivator implements. The H had dual front wheels, while the HN had a single front wheel. With their two-cylinder 99.7-cubic-inch engines, the Model Hs were rated at 12.5 drawbar horsepower and 14.8 belt horsepower. *Owner: Kenneth Anderson.*

1941 Model HWH
The Model HWH (wide and high) was manufactured for garden row-crop work and as a result had lightweight rubber tires, an adjustable wide front axle, and a minimum ground clearance of 21.4 inches. Only 126 were sold. *Owners: The Keller family.*

Model L

The styled Model L had a Hercules-built vertical twin 66.4-cubic-inch engine with a bore and stroke of 3.25x4 inches. The powerplant was rated at 9.1 drawbar horsepower and 10.4 belt horsepower at 1,550 rpm. The first production model had a smaller Hercules 56.5-cubic-inch engine. The tractor's three-speed gearbox allowed a top speed of 6 miles per hour. The L was manufactured at the Deere Moline (Illinois) Wagon Works between 1938 and 1941. *Owners: The Keller family.*

Model LA

The Model LA—produced from 1941 to 1946—was a big improvement on the L. While the Model LA was nearly 1,000 pounds heavier, it had a 77-cubic-inch Deere engine and produced 13.10 drawbar horsepower and 14.34 belt horsepower at 1,850 rpm. *Owners: The Keller family.*

Model M

The Model M—built from 1947 to 1952—was designed not only to replace the Models BR and BO, but also the H. To compete with the successful Ford-Ferguson 9N/2N, the M was fitted with the hydraulic Touch-O-Matic system for precise control of implements. More than 75,000 M and MT (tricycle) models were manufactured in the new Deere Dubuque, Iowa, factory. The Model M was a successful tractor with sales around 25,000 units per year, but Ford-Ferguson's 9N was more popular with a yearly production of 42,000. *Owner: Bob Pollock Collection.*

1950 Model R Diesel

Deere's replacement for its popular Model D—the Model R—came out in 1949 and was the company's first diesel tractor. When tested at Nebraska, the R's powerhouse 416-cubic-inch engine produced 45.70 horsepower at the drawbar and 51 horsepower on the belt running at 1,000 rpm. This Model R has spent its working life in Alberta, Canada. *Owner: Chris Dart.*

1953 Model R
The Model R used a two-cylinder horizontally opposed gasoline engine to get it started.
The pony motor had a cubic inch displacement of 25. *Owner: J. Newport.*

Chapter Four

The Numbered Series

1956 Model 70 Diesel Row-Crop

Introduced in September 1934, the Model 70 Diesel was Deere & Company's first diesel row-crop tractor. When tested at Nebraska, it set a new fuel economy record that lasted for years. The Model 70's 376-cubic-inch engine was considerably smaller than the 1949 Model R's powerplant (416 cid), but still produced essentially the same amount of power at the drawbar: 43 horsepower. Part of the reason for the engine's efficiency was the Model 70's weight: 6,035 pounds, 1,365 pounds (616.5 kg) fewer than the Model R.

1954 Model 40 High-Crop

Deere's tractor model naming system moved from being letter based to number based in 1952, when the Models 50 and 60 were introduced. The Model 40 was launched in 1953 to replace the utility Model M and had a 15 percent increase in power. The high-crop version pictured here could cultivate two rows and offered up to 32 inches of ground clearance. The tractor was also lengthened to improve stability. The Model 40 also came equipped with an excellent three-point hitch, live hydraulics, improved brakes, adjustable wheel tread, and power takeoff. *Owners: The Keller family.*

1954 Model 40 High-Crop
The vertical two-cylinder gas engine on the Model 40 had 100.5 cubic inches of displacement and produced 22.90 horsepower on the drawbar and 25.20 at the pto/belt at 1,850 rpm. The square engine had a bore and stroke of 4 inches.

1954 Model 40 High-Crop
At left, the modified height of the rear axle and mounting of the drawbar seem slightly crude for a production tractor. Above, the instrument panel with water and amp meter gages. Note the width of the engine cover, which has been designed for maximum visibility for for cultivation.

Model 50
The Model 50 was intended as a replacement for the Model B and was introduced in late 1952. One of the Model B's problems that the Model 50 intended to fix was that the B's right-hand cylinder always ran weaker, with less gasoline, than the left-hand cylinder. So the Model 50 came equipped with a Marvel-Schebler two-barrel carburetor that had a modified inlet manifold and cylinder heads. The carburetor gave better performance and economy. *Owner: Marv Mathiowetz.*

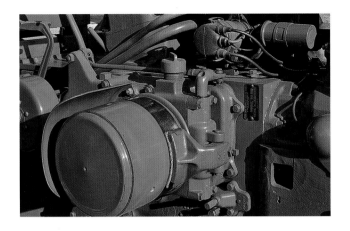

1955 Model 50
The power takeoff side of the Model 50's two-cylinder 190.4-cubic-inch gasoline engine. The tractor's pto/belt horsepower was rated at 28.55 when tested at Nebraska in 1952.

1955 Model 50 Row-Crop
The Model 50 was manufactured until 1956, with 32,574 coming off factory lines. More than 85 percent of those built were dual narrow-front row-crop tractors. For extra strength, the Model 50 used an angle iron frame rather than the pressed steel frame of the styled B. Note the shiny nonstandard chrome pipe instead of a muffler on this tractor. *Owner: Rich Ramminger.*

1953 Model 60
Introduced in June 1952 to replace the Model A, the Model 60 stayed in production until 1956. Some orchard and LPG Model 60s were manufactured in 1957. The tractor's 321-cubic-inch gasoline engine produced 36.90 drawbar horsepower and 41.60 pto/belt horsepower. Note the small bulge on the outside of the engine to protect the spark plugs from knocks. *Owner: Marv Mathiowetz.*

1956 Model 60 Row-Crop
The Model 60 came in orchard, standard-tread, row-crop, and high-crop versions. It also could be ordered with a gasoline, all-fuel, or LPG engine. *Owner: Rich Ramminger.*

1956 Model 70 Diesel Row-Crop
The Model 70's two-cylinder 376-cubic-inch engine produced 45.7 drawbar horsepower and 51.5 pto/belt horsepower. It had a bore and stroke of 6.125x6.375 inches. To start the diesel, a 12-horsepower V4 gasoline pony motor was used. The exhaust heat from the 18.8-cubic-inch pony engine was directed through the main diesel engine to help in starting. *Owner: Rich Ramminger.*

1956 Model 70 Diesel Row-Crop
With optional power steering, the Model 70 came with a three-spoke steering wheel. Models without power steering had four-spoke wheels. All Model 70s came with a comfortable deep-cushion seat and backrest, standard live power takeoff, and a separate clutch.

Model 70 Wide-Front High-Crop
The Model 70's Dual Touch-O-Matic hydraulic controls accommodate separate or coordinated lowering and raising of the front and rear implements.

1955 Model 80

The Model 80 was launched in 1955 as a replacement for the Model R. Its 472-cubic-inch diesel engine produced 61.80 horsepower at the drawbar and 67.60 horsepower at the belt. The Model 80's pony motor was the same as that on the Model 70: a 12-horsepower V4 gas powerplant. The Model 80 came with a six-forward gear transmission. *Owner: Don Miller.*

Model 80 Diesel

Only 3,485 Model 80s were manufactured in John Deere's Waterloo, Iowa, plant between 1955 and 1956. *Owner: Benjamin Bone.*

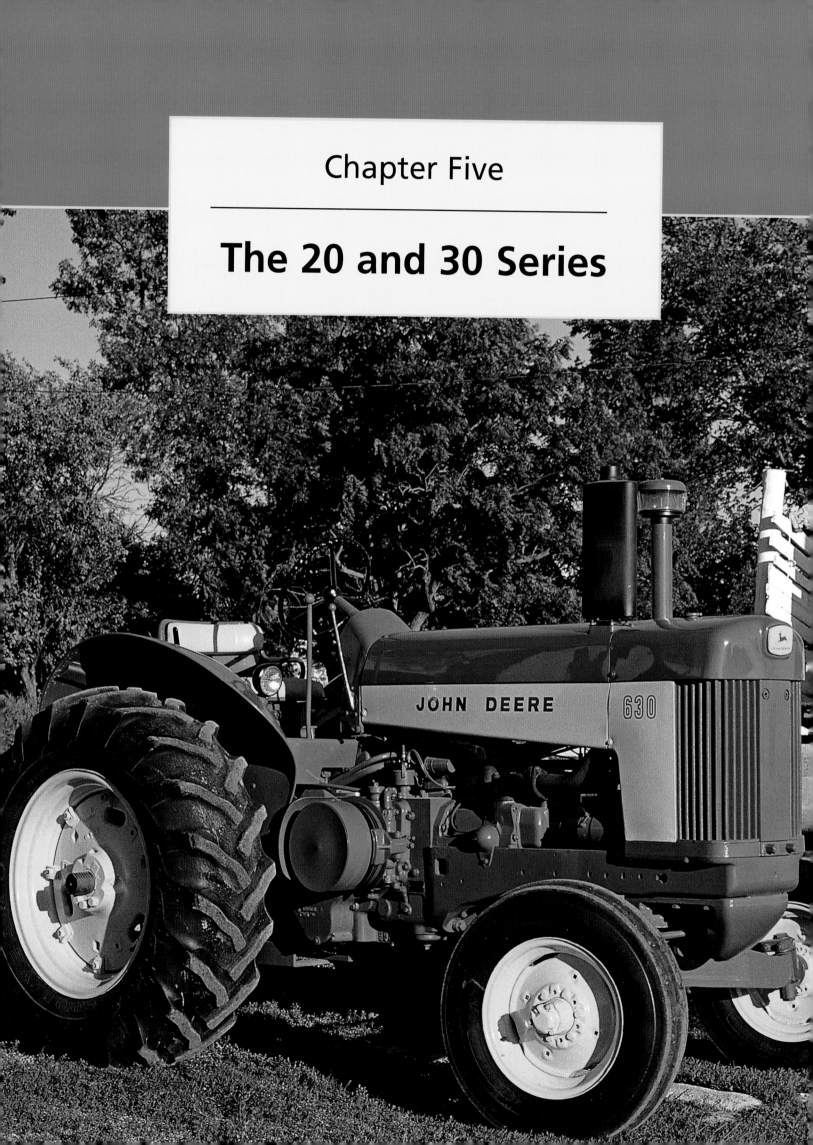

Chapter Five

The 20 and 30 Series

**1960 Model 630 and
1960 Model 435 Diesel**
The 30 Series tractors were the
last of Deere's two-cylinder
tractors and were known for
their operator comfort and
usability. In 1958, the 630, a
four-plow tractor, replaced the
620. By the end of 1960,
about 18,000 had rolled off
production lines. The Model
435 Diesel was manufactured
by pairing a GM Diesel 2-53
engine to a Model 430 utility
row-crop tractor.

Model 320 Standard

Derived from the Model H and Model M, the 320 was designed especially for the small farmer and vegetable grower. It came in two versions: the 320S, or Standard, and the 320U, or Utility. The 320 was powered by a vertical two-cylinder 22.4-horsepower gasoline engine with 100.5 cubic inches of displacement. This 320 was photographed at the Two-Cylinder Club show at the John Deere Historic Site in Grand Detour, Illinois.

Model 320
Only 3,084 Model 320s were manufactured between 1956 and 1958 at Deere's Dubuque, Iowa, factory. The early production 320s had a vertical steering wheel. This model has the later angled steering wheel.

Model 820 Diesel
Built from 1956 to 1958, the Deere 20 Series tractors had an attractive new look, more horsepower, and Custom Powr-Trol—an improved version of the dual-line hydraulic system. The largest tractor in the line, the Model 820, had a 472-cubic-inch two-cylinder engine, rated at 62 drawbar horsepower and 67.6 pto/belt horsepower. After the first 3,100 units were produced, the Model 820's engine was modified to boost power by 12 percent to 69.66 horsepower at the drawbar and 75.60 horsepower at the belt.

Model 420V

The 420—manufactured between 1956 and 1958—sold well, with approximately 46,450 coming off factory lines. Shown here is a rare 420V (Special) with 26-inch rear-axle clearance designed for vegetable farmers. The 420 came standard with a four-speed transmission, but a five-speed was also available. *Owner: The Pollock Collection.*

Model 420
The Model 420's engine had a 20 percent power increase from the Model 40's, which came as a result of increasing the bore by 0.25 inches to 4.25 inches. This modification gave the vertical two-cylinder powerplant a capacity of 114.75 cubic inches of displacement. The gasoline engine was rated at 27.08 drawbar horsepower and 29.21 pto/belt horsepower.

Model 520 Row-Crop
In a preproduction Nebraska test, the Model 520's two-cylinder 189.8-cubic-inch engine produced 36.11 pto/belt horsepower. This 520 has a single front wheel and high-clearance, 42-inch rear wheels and an extra-wide axle shaft. *Owners: The Keller family.*

Model 520
The Model 520's dual Custom Powr-Trol hydraulics and load-compensating three-point hitch.
The tire pressures visible above are on transfers on the side of the Float-Ride seat.

Model 520 LPG
The LPG-powered 520 had a
189.8-cubic-inch engine that
produced 34.31 horsepower
at the drawbar and 38.58
horsepower at the pto/belt.
The engine had a high-
compression ratio of 8.75:1.
The Pollock Collection.

Model 520 LPG
Controls and instruments on
the 520 LPG model.

1957 Model 620

The row-crop 620 was manufactured from 1956 to 1960 as the successor to the Model 60. Its power had been boosted 20 percent by increasing the engine speed from 975 to 1,125 rpm, producing 48.68 horsepower at the pto/belt. *Owner: Bruce Copper.*

Model 330
The Model 330 and 430 had vertical two-cylinder engines, the 330's capable of 100.5 cubic inches of displacement. It was rated at 21.5 drawbar horsepower and 24.9 pto/belt horsepower. Only 1,091 of the rare 330 were manufactured between 1958 and 1960.
Owner: Don Wolf.

1959 Model 430W
The Model 430 was available in so many customized options that some dealers starting referring to it as a "fleet." This is a 430W (row-crop utility), but the model was also offered as the 430S (standard), 430U (utility), 430H (high-crop), 430V (special), 430 F-3 (forklift), and 430C (crawler). The 430 had a 113.3-cubic-inch engine rated at 27.1 drawbar horsepower and 29.21 belt horsepower. *Owner: Don Wolf.*

Model 430T Row-Crop

The Model 430T was available with dual tricycle wheels, a single front wheel, or as a wide-front row-crop. Manufactured at Dubuque, Iowa, more than 3,250 Model 430Ts came off factory lines. The total production for the 430 was 12,680 units. This row-crop has the Custom Powr-Trol three-point hitch system. *Owner: Kelsi Ver Ploeg.*

Model 430U
Only 1,381 utility Model 430s were built at Dubuque, Iowa, between 1959 and 1960.

Model 430U
The utility 430 shown here has the regular fenders fitted. At the controls is Duane Ver Ploeg, who restored it for the owner, Jenna Ver Ploeg.

1960 Model 435 Diesel

Deere researched the possibility of designing its own diesel on the Model 435s, which would be developed from the vertical twin 113-cubic-inch gasoline engine used on the 430s, but that was not economically viable. The General Motors two-cycle 106-cubic-inch (1,736cc) two-cylinder supercharged diesel engine that powered the 435s was rated at 32.91 pto/drawbar horsepower. About 4,625 Model 435s were manufactured in 1959. *Owner: Don Wolf.*

1959 Model 530
The final version of the series that started with the venerable Model B, the 530 was a modern general-purpose tractor. Its insides had not changed much from the 520's, but the 530 had dual headlights as opposed to the 520's singles, a much different font used for the model number, and more yellow sheet metal along the engine. *Owner: Kent Bates.*

1960 Model 630
This 1960 630 has a two-cylinder 321-cubic-inch (5,258cc) 49-horsepower engine. *Owner: Larry Maasdam.*

Model 630 LPG

The general-purpose Model 630 row-crop had four front axle options: dual narrow front, Roll-O-Matic, wide front, and single wheel. The standard-tread 630 shown here is also a high-crop version. There was no 630 orchard tractor offered because the 620-O stayed in production. *Owners: The Keller family.*

Model 630 LPG Standard
This LPG 630 has a 302.9-cubic-inch high-compression (8.1:1) two-cylinder engine, rated at 45.8 drawbar horsepower and 50.3 pto/belt horsepower. Options fitted to this 630 include power steering and a Float-Ride seat with padded arm rests.

Model 730 Standard

The five-plow Model 730 was manufactured at Waterloo, Iowa, between 1958 and 1961, and 29,713 were built. When production finished, the tractor's tooling was shipped to Argentina, where production continued up to 1968 with about 20,000 produced, all electric-start diesels. Between 1958 and 1961, kit production of the 730 began in Monterrey, Mexico. *Owner: Robert Couch.*

Model 730 Standard
The Model 730's Float-Ride seat can be adjusted for different drivers' weights. Note the hydraulic reservoir for the Custom-Powr-Trol.

1959 Model 730
Introduced as an alternative to the pony motor, a 24-volt electric starter was a popular option on Model 730 diesel tractors near the end of the model's production life. *Owner: Lyle Pals.*

1959 Model 730 Diesel
The information transfer on the Float-Ride seat explains the correct settings for different drivers' weight.

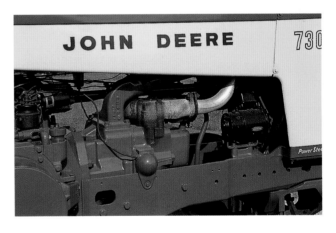

1959 Model 730 Diesel
The tractor's two-cylinder 360.5-cubic-inch gasoline engine puts out 53.05 drawbar horsepower and 59.12 pto/belt horsepower. The transmission has six speeds. The tractor's two-cylinder 360.5-cubic-inch gasoline engine puts out 53.05 drawbar horsepower and 59.12 pto/belt horsepower. The transmission has six speeds.

Model 830 Standard

Only offered in standard tread, the Model 830 was a six-plow tractor known as the Wheatland and manufactured in Waterloo, Iowa, between 1958 and 1960. Deere also offered a rice special version with mud shields and mud covers on the brakes. In all, approximately 6,715 Model 830s came off factory lines. *Owner: Robert Couch.*

Model 830 Engine
The 8,667-pound Model 830 was powered by a big two-cylinder diesel with 471.5 cubic inches of displacement. It was rated at 75.6 pto/belt horsepower at 1,125 rpm.

Model 830 Diesel
The 830's new improved controls and instruments, optional power steering, and Float-Ride seat with six-speed transmission, offering speeds from 2.33 to 12.25 miles per hour.

1959 Lanz 2816, 1952 Model AWH, and GMW 35
From left, a Mannheim-built Lanz D2816 in full John Deere colors and logo with a 28-horsepower diesel engine and a Deere AWH, the rare wide-front high-crop model.
Owner: Larry Maasdam. Inset is a GMW 35 row-crop, manufactured in Almhult, Sweden.

GMW 35 Row-Crop
The GMW 35 tractor with its horizontal twin-cylinder 5.3-liter engine was a copy of the late-model AW and was sold overseas. It was manufactured in Almhult, Sweden, by the Gnosjo Mekanski Werkstad Company in the 1950s. The company also built about 200 GMW 25s with a 2.9-liter engine, which was a copy of the BW. *Owners: The Layher family.*

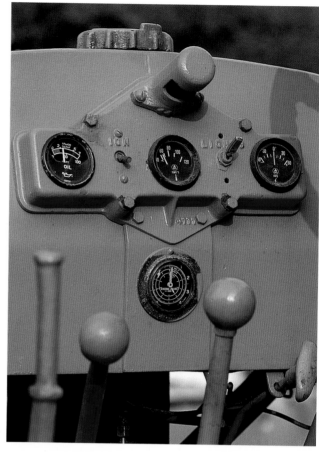

1950s Chamberlain 40K

Tractor mechanic Bob Chamberlain from Western Australia started designing and building tractors for Australian farmers in the late 1930s, but stopped during World War II. During the war, Chamberlain went to work in the United States designing tanks, but he returned to Australia in 1945 and developed the 40K at the family piston factory in Melbourne. Tractor production started in Perth, Western Australia, in 1949. The 40K was produced until 1955, with 2,000 built. The tractor had only a simple drawbar and no hydraulics (they were not needed by Australian farmers). The 40K's two-cylinder horizontally opposed kerosene engine was rated at 30.8 drawbar horsepower and 40 belt horsepower. The nine-speed transmission gave the tractor a top speed of 15 miles per hour. Deere & Company became a primary shareholder in Chamberlain in 1970 and bought out the company in the 1980s. *Owner: Aubrey Sanders.*

1950s Chamberlain Super 70 Diesel

The Chamberlain Super 70 was built from 1954 to 1963 with a three-cylinder supercharged two-stroke General Motors diesel engine. Maximum power was 77 belt horsepower, but the tractor was rated at 65.4 horsepower at the belt and 52.5 at the drawbar. Top speed was 18.75 miles per hour. The Super 70's transmission had nine forward gears and three reverse gears, using a three-speed range of low, medium, and high. Its clutch was hand operated. The tractor's leaf-spring front suspension improved handling and operator comfort.
Owner: Denzil Horler.

1950s Chamberlain Super 70 Diesel
The Chamberlain Super 70 engine's super-charger.

1965 Chamberlain Super 90

Chamberlain tractors were first painted yellow in 1975 after Deere purchased a controlling interest in the Australian tractor maker. The last of the yellow Chamberlains came off the production line in 1984. From then until the middle of 1986, when Chamberlain production ended, the tractors were painted John Deere green and built at the Welshpool factory in Western Australia. Now the Chamberlain name is gone, and the Deere tractors in Australia are imported from Mannheim, Germany, and Waterloo, Iowa. *Owner: John Smith.*

1939 Lanz HR9 Eil Bulldog

In 1956, John Deere acquired Lanz, a German manufacturer best known for its Lanz Bulldog tractor. With the purchase, Deere gained a European manufacturing site (at the time it was the only U.S. tractor producer without a European factory). This Lanz HR9 Eil Bulldog came equipped with a 629-cubic-inch single-cylinder semi-diesel engine. The road Bulldog models had five or six forward gears that could be changed on the move and a top speed of 25 miles per hour. *Owner: Pierre Bouillé.*

1939 Lanz Bulldog 9506

The Lanz Bulldog was the world's first hot bulb–fired tractor burning inexpensive crude oil. The 9506 model had a hot-bulb semi-diesel engine with 629 cubic inches of displacement. It was rated at 45 drawbar horsepower at 630 rpm. *Owner: Rik Van Esser.*

1948 Lanz Bulldog Allzweck 7506

Heinrich Lanz established his famous agricultural manufacturing company in Mannheim, Germany, in 1859, focusing on building steam engines and threshers. By 1911, he had started building his first tractor, the Landbaumotor. In 1921, Dr. Fritz Huber designed the hot-bulb semi-diesel Bulldog. This 7506's single-cylinder hot-bulb semi-diesel engine was rated at 25 horsepower. The Allzweck model came with rubber tires and a six-forward-speed gearbox. *Owner: Jaak Langers.*

1959 Lanz 2816
The 28-horsepower single-cylinder diesel engine on the Lanz 2816 was a full diesel compared with the semi-diesel of the earlier Lanz model. The semi-diesel used heat from a flame to start. The 2816 model was the last tractor in the Lanz series and was still built after Deere took over the company. *Owner: Jaak Langers.*

1959 Lanz 2816
Following Deere's purchase of Lanz in 1956, the Model 2816 bore the full John Deere Lanz name and colors. The transmission had six forward gears and one reverse gear. *Owner: Larry Maasdam.*

1951 Lanz Bulldog D9506

The Lanz D9506 *Ackerluf*, meaning *field air*, tractors were fitted with pneumatic tires. This model has the 629-cubic-inch (10.3-liter) hot-bulb semi-diesel engine, producing 45 horsepower. *Owner: Daniel Binet.*

1954 Lanz Bulldog D2806
The Lanz D2806 came with a 226-cubic-inch (3.7-liter) single-cylinder semi-diesel engine that produced 28 horsepower. *Owner: Daniel Binet.*

Chapter Seven

The New Generation of Power

Model 2510 Diesel High-Crop
Deere formally launched its New Generation models (all with four- and six-cylinder power-plants) in dramatic fashion on August 30, 1960, when 5,000 Deere dealers, industry leaders, and company personnel saw them chug across the floor of the Dallas Memorial Auditorium. The 2510 was last of the New Generation 10 Series, launched in the summer of 1965.

JOHN DEERE

2510
DIESEL

1959 Model 8010

The big Model 8010 was not only one of the first articulated four-wheel-drive tractors produced, but for Deere it was the first tractor in 40 years to have more than two cylinders. The 10-ton (9,000-kg) 8010 had a six-cylinder supercharged, two-cycle diesel engine rated at 215 horsepower. Built by General Motors, the engine was known as the GM 6-71.

1959 Model 8010

The 8010 shown here is serial number 1000, the first model built. It was never sold as a new tractor but was used by Deere & Company for tests and demonstrations. *Owners: The Keller family.*

1964 Model 1010
Grove-Orchard

The Dubuque, Iowa–built 35-horsepower 1010 was the smallest of the New Generation Deere tractors. This orchard version is one of 72 made and one of only 63 with a 115.5-cubic-inch gasoline engine. *Owners: The Keller family.*

Model 3010 Diesel Row-Crop

The Model 3010 also was built at Deere's Waterloo, Iowa, factory and came equipped with many new features. One of these was the Synchro-Range transmission, which could change gears whether going forward or backward without stopping. The transmission had eight forward gears and two reverse gears. Another New Generation innovation was the 3010's closed-center hydraulic system. It had three circuits: one for power steering, one for brakes, and one for the three-point hitch and remote implements. The 3010 also had a comfortable sprung deluxe seat, and its three-point hitch now had a Quick-Coupler option, allowing the center link to be connected without leaving the seat.

Model 3010 Diesel Row-Crop
This Model 3010 had the four-cylinder 254-cubic-inch diesel engine, which in 1960 Nebraska tests produced 52.8 drawbar horsepower and 59.44 pto/belt horsepower at 2,200 rpm.

1960 Model 3010 Row-Crop Utility
This Model 3010 row-crop has optional adjustable-tread axles. It is powered by a four-cylinder 201-cubic-inch gasoline engine. When tested at Nebraska, the engine produced 55.10 pto/belt horsepower. *Owners: The Keller family.*

1964 Model 5010
Weighing 13,400 pounds, the Model 5010 was powered by a six-cylinder 531.6-cubic-inch diesel engine that put out 106 horsepower at the drawbar and 121 horsepower at the belt. The 5010 could reach a top speed of 20 miles per hour. *Owner: Neil Bailey.*

1964 Model 5010
Owner, restorer, and farmer
Neil Bailey at the controls of
his very original 5010. Bailey
uses only John Deere tractors
for working his farm seven
days a week.

1964 Model 5010
The enormous six-cylinder diesel
on the Model 5010 has a bore
and stroke of 4.75x5.00 inches
(118.75x125mm).

Model 5010
A Model 5010 standard-tread tractor manufactured at Waterloo, Iowa, between 1963 and 1965.
Owner: Philip Coles.

1966 Model 2510 Diesel Row-Crop
The Model 2510 was the last of the New Generation 10 Series, launched in the summer of 1965 shortly before the 2510 became the 2520. In Nebraska tests, its four-cylinder 180.43-cubic-inch diesel engine was rated at 50.66 pto/belt horsepower for the Power-Shift model and 54.96 for the Synchro-Range model. *Owner: Marv Mathiowetz.*

Model 2510 Diesel High-Crop
The 2510 paired the 3010 chassis with the 2020 engine. The tractor was offered with a gasoline or diesel engine, but no LPG option. This rare high-crop was an expensive option, costing $1,200 more on a $4,475 2510 diesel. *Owners: The Keller family.*

1968 Model 1020
Fitted with a post rammer for fence work, this Mannheim-built three-cylinder diesel engine tractor is still working four decades later. The 1020 was the first John Deere tractor sold by the Tincknell Company in South West England. *Owner: Nigel Hutchings.*

1970 Model 1020
The three-cylinder diesel engine on the Model 1020 had 152 cubic inches of displacement and a bore and stroke of 3.86x4.33 inches. In a 1966 Nebraska test, the 1020 tractor recorded 38.82 pto/belt horsepower and 33 drawbar horsepower. It had eight forward gears and four reverse gears.

1970 Model 1020
This Mannheim-built tractor has instruments showing 4,153 hours of use and speed in miles per hour.

1970 Model 1020

Built at Mannheim with European lighting designed for the British market, the 1020 was a popular tractor for small farms. The tractor had a comfortable sprung seat with its gear levers and a steering wheel within easy reach. *Owner: Neil Bailey.*

1973 Model 1120

The 1120 was manufactured in Mannheim, Germany, between 1967 and 1975. The 5,557-pound tractor was powered by a three-cylinder diesel engine rated at 45 pto/belt horsepower. The Model 1120's engine had a bore and stroke of 4.02x4.33 inches.

1973 Model 1120
At right, the business end of the 1120, with its three-point hitch with lower-link sensing and 540-rpm PTO drive. The transmission had eight forward gears and four reverse gears. *Owner: Kings Farm Machinery.*

1967 Model 2020 Orchard
This rare Model 2020 has the full citrus fenders and wire mesh covering the steering wheel. Only 21 were built. The four-cylinder 180.71-cubic-inch gasoline engine on this tractor produced 53.7 pto/belt horsepower. The long clutch lever and collar-shift levers seen here are for the tractor's eight-speed transmission. The 2020 Series was manufactured from 1965 to 1971. *Owners: The Keller family.*

Model 2020 Row-Crop Utility
No narrow-front Model 2020s were built, but the tractor's adjustable front axle provided good crop clearance. *Owner: Robert Heckendorf.*

1969 Model 2020 Diesel

From 1969 to 1972, the Model 2020 was manufactured in Mannheim, Germany, using a Deere diesel engine built in Saran, France. The 201.8-cubic-inch engine had a bore and stroke of 3.86x4.31 inches and produced 46 drawbar horsepower and 54 pto/belt horsepower at 2,500 rpm. The Model 2020 was also manufactured at Deere's Dubuque, Iowa, works from 1965 to 1972. *Owner: Neil Bailey.*

1969 Model 2020 Diesel
The Model 2020's three-point hitch and independent PTO. The transmission had eight forward gears and four reverse gears. *Owner: Neil Bailey.*

1969 Model 2020 Diesel
The Model 2020's instrument panel and gear levers for the high-low PTO ratio and gearbox. Power steering was standard.

1972 Model 2120
The 20 Series tractors featured good-sized foot plates to make mounting and dismounting easier.

1968 Model 2120
This 1968 model is fitted with a Duncan cab and has spent most of its working life in Holland. *Owner: Neil Bailey.*

1972 Model 2120 Diesel
The four-cylinder diesel
engine built at the John
Deere engine factory at
Saran, near Orleans in France.
The 219-cubic-inch diesel
engine with a bore and stroke
of 4.02x4.33 inches was rated
at 72 horsepower PS, the
European classification, or
the English rating SAE of 79
horsepower.

1972 Model 2120 Diesel
The 2120 was produced at Deere & Company's factory at Getafe, near Madrid, Spain. The New Generation styling was popular in Europe and carried over to the 30 Series in 1973. Above and on the opposite page is a look at 2120's styled shell fenders and Roll-Gard protection (a popular option in North America). *Owner: Ivor Grant.*

1967 Model 3020 Row-Crop
Throughout the Model 3020's production from 1964 to 1972, the tractor went through constant changes to keep up with stiff competition from other manufacturers. This model is equipped with a four-cylinder 227-cubic-inch gasoline engine, rated at 71 pto/belt horsepower. It also has a round cylindrical muffler. The 1968 model has an oval muffler and a larger 241-cubic-inch four-cylinder engine. *Owner: Arvin Busenitz.*

Model 3020 High-Crop
This popular high-crop Model 3020 is equipped with the Roll-Guard rollover protective structure and canopy. *Owners: The Keller family.*

Model 3020 Orchard
This orchard tractor has all the citrus fruit specifications with complete rear fenders and engine side curtains.

1969 Model 3120

The Model 3120 tractors were manufactured in Mannheim, Germany, and Getafe, Spain, but their six-cylinder engines were built in Saran, France. The Roll-Gard safety frames and optional canopy seen here were not initially popular options until John Deere promoted them with many farm safety groups. *Owner: Neil Bailey.*

1969 Model 3120

Note the Roll-Gard safety protection, three-point hitch, and power takeoff on the 3120. The rear fenders on this tractor are standard European style. A full European style was an optional extra. The full European-style fenders enclose the tire and stop mud and water from being spread over the driver and following vehicles. These are a legal requirement in Europe for highway use. *Owner: Neil Bailey.*

1969 Model 3120
The 3120 sold as a 95-horsepower SAE tractor in England, an 86PS-horsepower tractor in Europe, and had an 81-pto/belt horsepower rating in North America. At left is the characteristic radiator surround used on larger-engined 20 Series tractors. Its extra top mesh provides additional ventilation.

Model 4020 LPG High-Crop
This Model 4020 LPG high-crop is a rare tractor, one of only 17 built. The 4020 was the most popular tractor of the 1960s. Its design, build, and reliability were all solid. The 4020's single-lever Power Shift transmission held up to tough use, unlike the disastrous Ford Select-O-Speed. *Owners: The Keller family.*

Model 4020 LPG High-Crop
The 4020's six-cylinder LPG engine was rated at 90.48 pto/belt horsepower in the 1963 Nebraska test. When re-tested in 1966 with a longer stroke, the engine produced 94.57 pto/belt horsepower.

Model 4020 High-Crop LPG and Diesel
Two very rare 4020s. Only 17 high-crop LPG 4020s and 170 diesel versions were built. *Owners: The Keller family.*

Model 4020 Diesel High-Crop
In 1963, the 4020's six-cylinder 404-cubic-inch (6,618cc) diesel engine was rated at 91 horsepower. In 1969, the engine produced 96 horsepower as a result of six years of minor modifications. *Owners: The Keller family.*

Model 5020

The Model 5020 introduced in 1965 had 10 more horsepower than the 5010. Its six-cylinder 531.6-cubic-inch diesel engine was rated at 113.72 drawbar horsepower and 141.3 pto/belt horsepower in the 1969 Nebraska tests. The 5020's Synchro-Range transmission had eight forward gears and two reverse gears. The standard-tread 5020 was built in Waterloo, Iowa, until 1972.

1966 Model 5020

The 5020 came with power steering and dual remote hydraulics for controlling implements.

Model 5020 Row-Crop
With a shorter wheelbase
than the standard-tread 5020
and utility-type king pins, the
5020 row-crop had three
inches of more crop clearance
and width adjustment.
Owner: Brent Johnson.

1966 Model 5020 Diesel
The 5020's boost in power over the 5010 came from developments on the engine head and injectors, plus a small increase in compression ratio. The bore and stroke on the 5010 and 5020 engine was the same: 4.75x5 inches. This heavyweight 16,000-pound tractor was the first 5020 imported into Britain by John Deere dealer Drake & Fletcher Ltd. of Maidstone, England. *Owner: Ivor Grant.*

1966 Model 5020 Diesel
At 1,000 rpm, the Model 5020's pto/belt horsepower was rated at 133.25, and drawbar horsepower was 113.72 at 2,200 rpm.

Model 8020

Model 8020s were rebuilt 8010s that also had been relabeled. The 8010's nine-speed transmission was faulty, so it was replaced on the 8020 with a heavy-duty eight-speed transmission. Despite their rebuild, Model 8020s were slow sellers—not finding a market until 1965 (four years after being launched). Not many implements were available for the tractor, which was also expensive. Only about 100 Model 8020s were built. *Owner: Paul Kleiber.*

Model 8020

The Model 8020s came equipped with a six-cylinder 425-cubic-inch General Motors 6-71 supercharged engine. It produced 215 pto/belt horsepower. The transmission had eight forward gears and four-wheel-drive capability. *Owner: Paul Kleiber.*

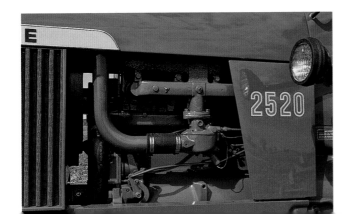

1969 Model 2520 High-Crop
This is one of only eight gasoline high-crop versions of the 2420. This tractor's boost in power over the 10 Series equivalent came from an increase in cubic inch displacement capacity: from 180 to 202. *Owners: The Keller family.*

1969 Model 4520 Row-Crop
The 4520 was Deere's first turbocharged tractor. While Allis Chalmers had been successful with the turbocharged D-19 back in 1961, Deere had resisted turbocharging for years. The advantage to adding a turbocharger was that it could boost engine power without adding much weight. The six-cylinder 404-cubic-inch turbocharged diesel engine on the 4520 was rated at 123.40 pto/belt horsepower. *Owners: The Keller family.*

Model 4320
Only offered as a row-crop tractor with an eight-speed Synchro-Range transmission, the Model 4320 was produced between 1971 and 1972 in Waterloo, Iowa. Its six-cylinder 404-cubic-inch turbocharged diesel engine was rated at 116.6 pto/belt horsepower in Nebraska tests. *Owners: The Keller family.*

1971 Model 4000 Low-Profile
This rare model, of which only 46 were manufactured, has the 3020's front axle and steering, but smaller rear tires with four-inch lower fenders. The 4000 was a cheaper version of the 4020. *Owners: The Keller family.*

Model 7520

Manufactured between 1972 and 1975, the Model 7520 was a hit in the specialized four-wheel-drive articulated tractor market. More than 2,000 sold. Powered by Deere's own six-cylinder 531-cubic-inch turbocharged and intercooled diesel engine, this tractor was rated at 165.2 drawbar horsepower and 175.80 pto/belt horsepower at 2,100 rpm. The transmission was an eight-speed Synchro-Range with high-low settings, giving 16 gears.

1968 Model 820

The Model 820 was manufactured in Mannheim, Germany, and was the smallest tractor in the John Deere range in 1968. Its three-cylinder 151.9-cubic-inch diesel engine had a 32 drawbar horsepower rating and a 31 pto/belt horsepower rating. It was popular in Europe as a two-bottom plow tractor because of its live-PTO mid-mounted takeoff for mowers. *Owner: Neil Bailey.*

1974 Model 920
The Mannheim-built 920 tractor uses the three-cylinder Deere diesel engine. *Owner: Neil Bailey.*

1974 Model 920
Note the three-point hitch, tow bar, and power takeoff with shield.

1974 Model 920
The instrument panel on the German-built Model 920 showing 8,534 hours of use. The speedometer is calibrated in kilometers per hour for each of the tractor's eight gears.

1974 Model 920

The Model 920 was also built in Mannheim, Germany, and had a 151.9-cubic-inch engine rated at 40 pto/belt horsepower. The tractor was designed so that farmers could have easy access to all the components, especially the dipstick and oil filter. This 920 has the American-type shell fenders. *Owner: Neil Bailey.*

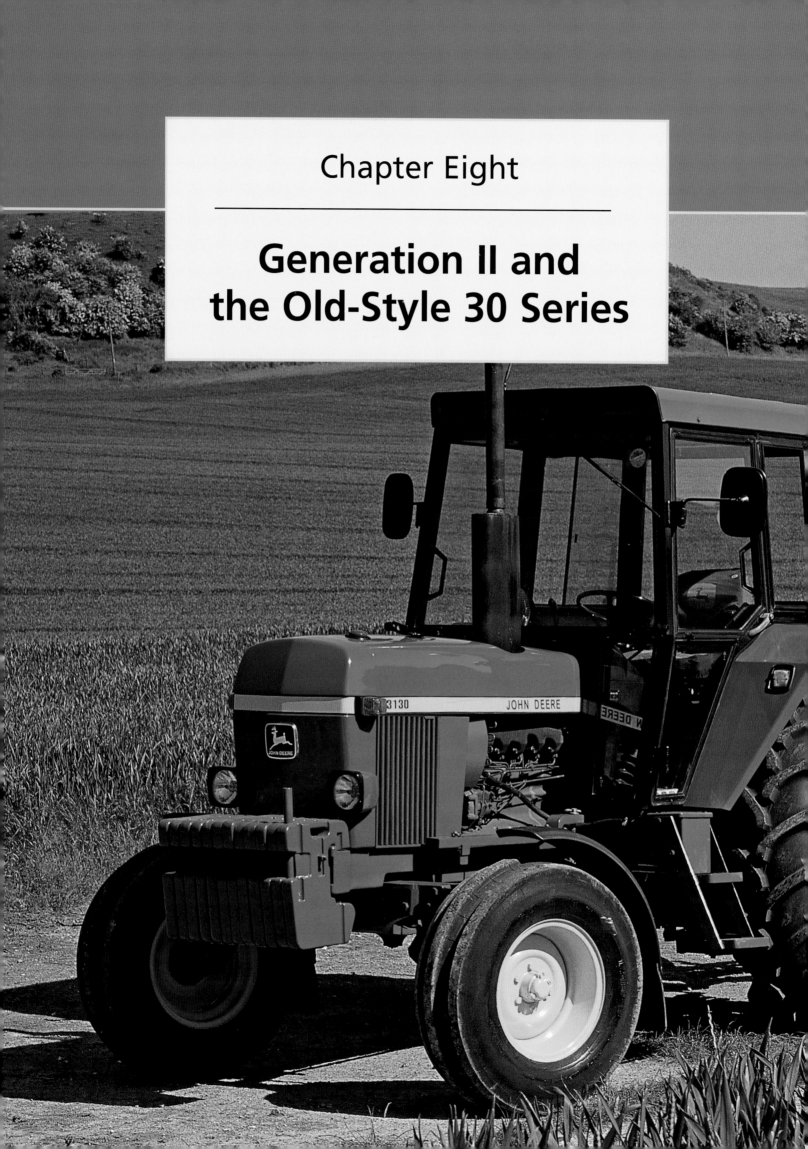

Chapter Eight

Generation II and the Old-Style 30 Series

1979 Model 3130
Deere & Company began launching its Generation II tractors in 1974, with the U.S. versions sporting a new look. The next year the tractors rolling out of the Mannheim, Germany, plant also had an updated appearance. The Model 3130, shown here, was the first fitted with the OPU safety-cab option. By 1979, a 12-speed high/low transmission (inset) was standard on the Generation II tractors. *Owner: Tony Adams.*

1975 Model 2030

This is one of the last old-styled 30 Series tractors Deere manufactured. Built in Mannheim, Germany, in 1975, this 2030 had a four-cylinder 219-cubic-inch diesel engine with a bore and stroke of 4.02x4.33 inches. It was rated at 60.6 pto/belt horsepower at 2,500 rpm. The 2020 was manufactured in Dubuque, Iowa, for the first couple years of production, before models started coming out of the Mannheim works in 1974. The Dubuque and Manheim models were very different, and in Canada the 2030 was sold as an 1830. *Owner: Nigel Hutchings.*

1975 Model 2030
The Model 2030's three-point hitch, power takeoff, and tow bar. This 1975 model also has an eight-speed transmission.

Model 2130
This Model 2130 came equipped with the optional factory-fitted safety cab. This tractor was manufactured between 1975 and 1980 with a four-cylinder 79PS-horsepower diesel engine.

Model 3130
Old-style New Generation Model 3130s were manufactured in Mannheim from 1972 to 1974 with six-cylinder 329-cubic-inch diesel engines rated at 80.65 pto/belt horsepower. *Owner: Nigel Hutchings.*

1979 Model 3130
This Model 3130 is mated with a 1970s-era Deere four plow.

1979 Model 3130
The 1979 Model 3130 had a six-cylinder 329-cubic-inch diesel engine that produced 97 drawbar horsepower at 2,500 rpm. Its bore and stroke was 4.02x4.33 inches.

1979 Model 3130
The Model 3130's three-point hydraulic hitch, power takeoff, and tow bar.

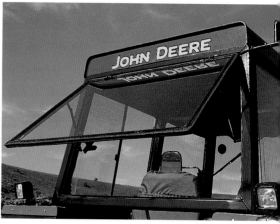

1979 Model 3130

By 1979, the restyled European 30 Series matched the
Waterloo-built Deere tractors. Above and on the opposite
page is a Generation II 3130 with OPU (operators protection
unit) cab, which was manufactured by Secura in their British
and Danish factories and shipped to Mannheim. This popular
cab was fitted from 1974 until 1981 when the SG2 Sound-Gard
cab replaced it.

Model 4030
Manufactured between late 1972 and 1975, the Model 4030 was one of the first Deere tractors equipped with the new Sound-Gard cab. The cab reduced noise by about 10 decibels over the previous OPU cab. The 4030 had a six-cylinder 329-cubic-inch (5,389cc) engine rated at 80 horsepower. *Owner: Glen Braun.*

1972 Model 4430
The popular six-cylinder 404-cubic-inch (6,617cc) turbocharged diesel on the Model 4430 gave the tractor a good power-to-weight ratio. The engine was rated at 126 horsepower and weighed 11,350 pounds.

Model 4630
Manufactured between 1973 and 1977, the Model 4630 had the same turbocharged engine as the 4430, but was fitted with an intercooler. The six-cylinder 404-cubic-inch diesel produced 135 drawbar horsepower and 150 pto/belt horsepower.

Model 8430

Manufactured between 1975 and 1978, the Model 8430 was an articulated four-wheel-drive tractor powered by a six-cylinder 466-cubic-inch (7.6-liter) turbocharged diesel engine. The powerplant produced 160.4 drawbar horsepower and 178 pto/belt horsepower. The tractor's transmission was a 16-speed Quad Range. *Owner: Mark Farwell. (Opposite page)/Owners: The Chaddock family (above, right).*

Model 8630
Big brother to the 8430, the 8630 had a six-cylinder 619-cubic-inch (10-liter) turbo diesel engine that produced 225 drawbar horsepower. The tractor's hydraulic Quik-Coupler hitch was designed for Category 2 and 3 implements.

Model 8640

The Model 8640 was a Category 3-N tractor manufactured between 1978 and 1982. Its six-cylinder turbocharged and intercooled engine had 619 cubic inches of displacement and was rated at 208.4 drawbar horsepower and 228.8 pto/belt horsepower.

1980 Model 4440

The Model 4400 was part of Deere's "Iron Horses" 40 Series tractor line, announced in late 1977. Its six-cylinder 466-cubic-inch turbocharged diesel engine put out 130 drawbar horsepower at 2,200 rpm. This two-wheel-drive 4440 with a Sound-Gard cab had a three-point hitch, 540 power takeoff, and an uprated hydraulic capacity for more lifting ability.

Model 4440
The levers for the 4440's 16-speed Quad Range transmission and Power-Assist Park Release.

1980 Model 8440 Articulated Four-Wheel-Drive
This Model 8440 (above and on opposite page, bottom) is powered by a 466-cubic-inch (7,633cc) turbocharged and intercooled six-cylinder diesel engine, rated at 180 horsepower. *Owner: Don Bray.*

Chapter Nine

New Power and the Modern Era

Model 6910 and Model 9300
The Model 6910 was among the best-selling tractors in Europe in the late 1990s because it could easily travel from farm to farm. Inset is a Model 9300, part of the 9000 Series, the largest tractors ever built by Deere when they were introduced in 1996. *Owners: S. Gray & Sons.*

Model 1040
Produced between 1979 and
1986, the new-syle 40 Series
included five different
models. In Europe, these
were the 1040, 1140, 2040,
2140, and 3140. The U.S.
equivalents were the 2040,
2240, 2440, 2640, and 2940.
The 1040, seen here in New
Zealand's Marlborough
winery area, was rated at 44
drawbar horsepower.

Model 1050
The Model 1050 was introduced in 1980 as one of the Deere's newest compact tractors. Manufactured by Yanmar in Japan, the 1050 had a three-cylinder 105-cubic-inch turbocharged diesel engine. The powerplant was rated at 33.4 pto/belt horsepower. This 1050 spends its days in Kaikoura, New Zealand, hauling in fishing boats.

Model 4050
Manufactured between 1982 and 1986, the Model 4050 became one of the quietest tractors ever tested at Nebraska when it only reached 70 decibels of sound at 50 percent load. The tractor's six-cylinder 466-cubic-inch turbocharged diesel was rated at 100 pto/belt horsepower. *Owner: Neil Bailey.*

Model 4050
This Model 4050 with attached hedge trimmer is equipped with the optional Caster/Action mechanical front wheel drive, which gives the tractor a small turning radius.

Model 4650
This Model 4650 carries two front-mounted sprayer tanks and a rear planter. It is powered by six-cylinder turbocharged and intercooled diesel engine with 466 cubic inches (7,633cc) of displacement. The engine is rated at 165.7 pto/belt horsepower. *Owner: Donald Schaefer.*

1996 Model 6400
The Model 6400 was the top-of-the-line four-cylinder tractor in Deere's 6000 Series. Its 276-cubic-inch (4521cc) diesel produced 100 horsepower. The Model 6400 here is pulling a Model 545 round baler in Somerset, England.

Model 6900

Introduced in 1994, the Model 6900 four-wheel-drive tractor was part of the 6000 Series and came equipped with a six-cylinder 414-cubic-inch (6.7-liter) turbocharged engine. The powerplant produced 130 belt/pto horsepower. The 6900 at right is pulling a Klaas Quadrant 1150 baler and sledge on the Somerset Levels in England. Below, a 6900 is equipped with front weights and a reversible John Deere six plow.

Model 7700
Manufactured between 1992 and 1996, the Model 7700 was one of the most powerful tractors in the 7000 Series. Its six-cylinder turbocharged/intercooled engine had 466 cubic inches of displacement and was rated at 126 pto/belt horsepower. This 7770 is mated to a seven-bottom reversible plow and is competing in a plowing competition in Somerset, England.

Model 7710 and Massey Ferguson Combine
Introduced in 1996, the Model 7710 was updated from its 7000 Series equivalent (the 7700) with a new PowerTech engine. The six-cylinder 497-cubic-inch (8.1-liter) diesel powerplant produced 138 horsepower. The 7710's electronic diesel injection system cleaned up the tractor's emissions and black smoke, and it also improved economy.

1992 Model 7800

Equipped with a new and improved ComfortGard cab, the Model 7800 only subjected operators to 72.5 decibels of noise. Its six-cylinder, turbocharged, intercooled, and 466-cubic-inch (7.6-liter) engine produced 135.6 drawbar horsepower and 146.7 pto/belt horsepower. The cutaway (below, right) shows the engine's internal workings, the tractor's 19-speed Power Shift transmission, and its four-wheel drive.

1984 Model 8850
The largest of the 8000 Series, the Model 8850 weighed 37,700 pounds. This immaculate giant Deere tractor has never suffered any problems or breakdowns in more than 6,000 hours of work. *Owner: Ruth Schaefer.*

Model 8850

The four-wheel-drive Model 8850 set a new power benchmark for Deere when its V-8 turbocharged/intercooled diesel engine was rated at 207 drawbar horsepower and 304 pto/belt horsepower. The massive powerplant had 955 cubic inches of displacement. The tractor had a six-speed Quad-Range transmission.

1996 Model 9300
The Model 9300, a four-wheel-drive articulated tractor, came equipped with a mammoth four-valve-per-cylinder 767-cubic-inch Powertech engine. It could produce 360 horsepower. The 9300 also came equipped with a Command View cab with side door entrance. The machine's hitch hydraulics and power controls were available at the operator's fingertips. *Owner: Ruth Shaefer.*

2001 Model 6310SE

The SE model of this Deere tractor was about 10 percent cheaper than the standard specification 6310. Yet with a 276-cubic-inch (4.5-liter) 99-drawbar-horsepower engine, the 6310SE was still a mighty machine. Its PowerQuad transmission had 16 gears.

Model 6910 and Model 6850 Combine

The Model 6910 came equipped with a six-cylinder turbocharged/intercooled 417-cubic-inch (6.8-liter) engine. The powerplant produced 144 pto/belt horsepower at 2,100 rpm. The tractor's 20-speed Autoguard transmission could travel speeds between 1.6 to 25.6 miles per hour. *Owners: S. Gray & Sons.*

Model 9420

This 2002 Model 9420 is powered by a six-cylinder 765-cubic-inch (12.5-liter) diesel engine that produces 425 drawbar horsepower and 302 pto/belt horsepower. The transmission options on 9000 Series tractors included a 12-speed Power Shift or Synchro-Range, or a 24-speed PowrSync.

Model 5320

The two-wheel-drive Model 5320 is among the many small utility garden tractors Deere & Company manufactures in Augusta, Georgia. It has a three-cylinder 179-cubic-inch (2.9-liter) diesel engine that is rated at 64 drawbar horsepower and 56 pto/belt horsepower. The tractor shown here has an optional 12-speed PowrReverser transmission.

Model 6620
These two Model 6620s are pulling New Holland 648 round balers. The tractor on the left is a 2005 model and on the right is a 2004 model with a Quicke Q980 loader. The 6620s are powered by a six-cylinder turbocharged/intercooled engine that produces 125 horsepower.

2005 Model 6620
The Model 6620 is seen here with a New Holland round baler, about to eject a tied round bale. The behemouth tractor is powered by a turbocharged intercooled six-cylinder 125-horsepower engine of 6788cc.

2004 Model 6620
Note the power drive to the baler from the rear PTO, which is rated at 540 rpm or 1,000 rpm.

Model 6620
Similar 2004 and 2005 model John Deere 6620s, both with New Holland 648 round balers. The two outfits belonging to contractor Mike Curtis can deal with a large acreage in a very short time, only stopping to reload cord for the round bales.

2004 Model 6820

This 6820 is powered by a six-cylinder turbocharged/intercooled 6,788cc engine rated at 135 drawbar horsepower at 2,100 rpm. It has a 20-speed Power Quad transmission and AutoGuard Plus with the Soft Shift feature (automatically compensating for loads). The tractor also could be ordered with a EcoShift or Auto Power transmission. The Auto Power transmission incorporates a clutch that is activated by the foot brake pedal.

Model 6920S

The 6920 tractor was the top model in the 6020 Series line with its PowerTech 4V-CR engine. The six-cylinder 160-horsepower powerplant has four valves per cylinder with centralized injectors and the Common Rail injection system.

2005 Model 6920

This Model 6920 is using a Krone twin-rotary side-delivery rake at silage time on the Somerset Level in England. The tractor also is fitted with optional turning front fenders, which improve the 6920's turning radius by 20 percent. Note the cylinders on the front axle for the hydro-pneumatic triple-link suspension system. *Owners: Malcolm Harding Contractors.*

Chapter Ten

The 2007 Models and 30 Series

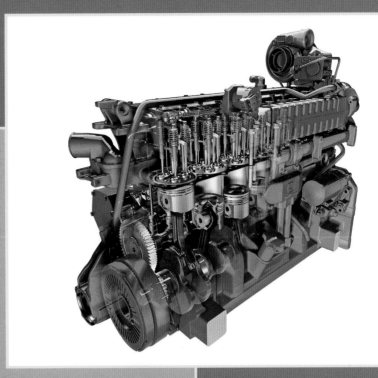

2007 Model 8430T and 9-Liter PowerTech Plus Engine

The Model 8430T showcases the new tractor design for the 2007 model year, with large ventilation louvers (very different from the 2006 panel louver vents). As a "tracked" tractor, the 8430T offers more power to the ground and outstanding ride and flotation on a variety of fields. It is especially useful on steep hills and slopes. Inset is a state-of-the-art tractor diesel engine. It features four valves per cylinder, a variable geometry turbocharger, intercooled and air-to-air aftercooled capabilities, EGR (exhaust gas recirculation), and a high-pressure common rail fuel system. This powerplant is rated between 225 horsepower in the 8130 and 330 horsepower in the 8530. It is the leader in its class in power torque and economy.

2007 Model 6330 and Model 633 Loader
Part of Deere's 6030 Premium Series Tractors line, the 2007 Model 6330 comes equipped with a four-cylinder turbocharged/intercooled 4,530cc engine. The powerplant is rated at 120 horsepower at the PTO and in transport mode.

2007 Model 6430
The 6430's engine features Deere's Intelligent Power Management System, which is tailored to farmers who use rear PTO in high-power operations or use their tractors as transport vehicles. With "intelligent power," the 6430 automatically can produce 10 more horsepower.

2007 Model 6830

This six-cylinder 150-horsepower tractor is equipped with GreenStar AutoTrac assisted steering. This system reduces overlapping field work by sending highly exact row locations to the tractor via RTK radio and then plotting the tractor's path via GPS. The RTK system has an accuracy of plus or minus two centimeters. With this type of steering, farmers can save time, labor costs, and fuel.

2007 Model 7430

This Model 7430 has a Model 753 loader structurally integrated into the tractor's chassis frame. The loader can precisely load and unload bales with the 7430's integrated joystick and left-hand reverser. The 7430's six-cylinder 6.8-liter PowerTech Plus engine is rated at 165 horsepower. With the Intelligent Power Management System, the 7430 can put out 190 horsepower for mobile PTO and transport on the road.

2007 Model 7730
The Model 7730 is also powered by the six-cylinder 6.8-liter PowerTech Plus engine, but is rated at 220 horsepower with the Intelligent Power Management System.

2007 Model 8530
The cutaway shows the 9.0-liter PowerTech Plus engine and parts of the transmission. This technically advanced diesel powerplant is capable of outstanding power, high torque, good fuel economy, and reduced emissions. It is rated at 330 horsepower with 360 horsepower available for road use and mobile PTO.

2006 Model 9520T
The Model 9520T is a top-of-the-line tracked tractor. It comes equipped with a six-cylinder 12.5-liter PowerTech diesel engine, rated at 450 horsepower. The Automatic PowerShift transmission has 18 forward speeds and 6 reverse gears. The tracks on this machine are offered in 30-inch (760mm) or 36-inch (900mm) widths, depending on a farm's needs.

2007 Model 8530

The 2007 Model 8530 is powered by a 9-liter 330-horsepower PowerTech Plus engine, which is Tier 3 emissions compliant (reducing nitrous oxide transmissions by 40 percent).

Index

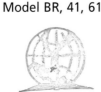